Sea Po

Compiled by John Foster

OXFORD

Oxford University Press, Great Clarendon Street, Oxford
OX2 6DP

Oxford New York
Athens Auckland Bangkok Bogota Bombay
Buenos Aires Calcutta Cape Town Dar es Salaam
Delhi Florence Hong Kong Istanbul Karachi
Kuala Lumpur Madras Madrid Melbourne
Mexico City Nairobi Paris Singapore
Taipei Tokyo Toronto

and associated companies in
Berlin Ibadan

Oxford is a trade mark of Oxford University Press

Published 1991
Reprinted 1992, 1996
ISBN 0 19 916424 X
Printed in Hong Kong

A CIP catalogue record for this book is available from the British Library.

Acknowledgements
The Editor and Publisher wish to thank the following who have kindly given permission for the use of copyright materials:

Finola Akister for 'My Castle' ©1990 Finola Akister; Moira Andrew for 'A Week of August Weather' ©1990 Moira Andrew; Stanley Cook for 'The Shell' ©1990 Stanley Cook, first published in *Word Houses*; John Foster for 'Sand' and 'The Lightship' both ©1990 John Foster; Theresa Heine for 'Kites' ©1990 Theresa Heine; Jean Kenward for 'Gulls' ©1990 Jean Kenward; John Kitching for 'Pebbles' ©1990 John Kitching; Judith Nicholls for 'To the sea' ©1990 Judith Nicholls; Irene Rawnsley for 'At the seaside' ©1990 Irene Rawnsley; Raymond Wilson for 'The Lighthouse' ©1990 Raymond Wilson.

Although every effort has been made to contact the owners of copyright material, a few have been impossible to trace, but if they contact the Publisher correct acknowledgement will be made in future editions.

Illustrations by
Rachel Ross, Katey Farrell, Bucket, Alan Marks, Jill Newton, Jill Barton.

Kites

Kites flying,
Swoop and sway,
Along the beach
On a windy day.

Kites sparkling
Like sea spray,
Along the beach
On a sunny day.

Kites flapping,
Wet and grey,
Along the beach
On a rainy day.

Kites tossing,
Up and away,
Along the beach
On a stormy day.

Theresa Heine

To the sea!

Who'll be first?
Shoes off,
in a row,
four legs fast,
two legs slow . . .
Ready now?
Off we go!
Tip-toe,
dip-a-toe,
heel and toe –
Yes or no?
Cold as snow!
All at once,
in we go!
One,
two,
three,

SPLASH!

Judith Nicholls

Sand

Sand in your fingernails
Sand between your toes
Sand in your earholes
Sand up your nose!

Sand in your sandwiches
Sand on your bananas
Sand in your bed at night
Sand in your pyjamas!

Sand in your sandals
Sand in your hair
Sand in your trousers
Sand everywhere!

John Foster

My Castle

It took hours to build my castle
In the sand, the other day.
The tide came in, and in a flash
Had washed it all away.

Finola Akister

At the seaside

I walked on the beach
in my brand new clothes
and stood on the sand
at the edge of the sea
and the sun was shining, shining.

I stepped to the water
on slippery stones
in my brand new shoes
to search for crabs
in the rock pools shining, shining.

A big wave spilled
and toppled me
in my brand new clothes
to the cold wet sea
with the pebbles shining, shining.

They brought me home
in my wet-through clothes
to my tucked-up bed
with a brand new cold
and a sore nose shining, shining.

Irene Rawnsley

A Week of August Weather

One Saturday when the sun was hot,
We set off for the sea,
My dad, my mum, old Herbert Bear,
My bucket, my spade – and me.

On Sunday a chilly breeze sprang up
And the sea looked cold and grey.
To keep ourselves warm we tramped the hills –
We went for a *very* long way!

On Monday the rain came pouring down;
We went out and got soaked through.
So we stayed inside all afternoon
And kept wondering what we could do.

On Tuesday morning the sun peeped out
So we raced down to the sea –
By afternoon it was raining once more,
Back indoors again long before tea!

On Wednesday a soft sea-mist rolled in
Making the shore a mysterious place,
And I was glad to hug old Herbert Bear
With his tattered familiar face.

On Thursday lightning lit up the sky
And a storm laced the waves with foam,
So we dodged about from shop to shop
Buying presents for people at home.

Then Friday dawned a *beautiful* day
So we paddled and soaked up the sun.
We picnicked, built castles, found dozens
Of shells and squeezed in a whole week of fun!

On Saturday the sun was still hot
When we waved good-bye to the sea,
My mum, my dad, old Herbert Bear,
My collection of shells – and me.

Moira Andrew

The Lighthouse

What I remember best about
my holiday was how, each night,
the lighthouse kept sweeping my bedroom
with its clean, cool ray of light.

I lay there, tucked up in the blankets,
and suddenly the lighthouse shone:
a switched on torch that stabbed the night
with its bright beam and moved on.

Then back it came, out of the dark,
and swung round, as in some fixed plan:
the light of the lighthouse – opening,
folding, and closing like a fan.

Raymond Wilson

The Lightship

The lightship guards the mouth of the bay
To warn other ships to keep away,
To steer clear of the rocky shore
Where many a ship has been wrecked before.

Through gales and storms, through day and night,
The lightship flashes its yellow light
Warning sailors to keep away
From the jagged rocks beneath the bay.

John Foster

Pebbles

I pull the pebble from the sand
And hold the pebble in my hand.
It feels so smooth and cool and old.
I dream the stories it's been told –
Of princes travelled from distant lands,
Of pirates roaming in fierce bands,
Of mighty fighting fish and whales,
Of savage storms and wrecking gales,
Of dead men drowned that tell no tales,
Of battleships, of love, of hate,
Of fisher-wives who wait and wait.
A million pebbles on the beach
And each its different tale to teach.

John Kitching

The Shell

In winter I put a shell to my ear
And through it I hear
The sound of the sea
Answer me.

'Are the donkey and funfair,
Boats and gulls still there?
The pier wading out from the land
And starfish like badges on the sand
Will they be there when I come next year?'
The whispering tide
In the shell replies,
'They will all be here
When you come next year.'

Stanley Cook

Gulls

I went to the sea shore,
I stood upon the sand,
My face to the water,
My back to the land.
I listened to the seagulls
I heard them call and cry,
They dipped to the water
And soared to the sky.

I went to the sea shore
And nobody was there
Except the tossing seagulls
That blew about the air.
I stood, and I watched them,
I heard them cry and call
With only me to answer
And no one else at all.

Jean Kenward